How to Control Electrical Noise

Michel Mardiguian

Interference Control Technologies, Inc.
Gainesville, Virginia

Interference Control Technologies, Inc.
Route 625, Gainesville, VA 22065
TEL: (703) 347-0030   FAX: (703) 347-5813

Library of Congress Catalog Card Number: 81-70305
ISBN: 0-932263-22-4

ACKNOWLEDGEMENT

The author wishes to thank Donald R.J. White who encouraged him to write this book on *How To Control Electrical Noise*. He expresses his appreciation to the individuals and companies who have furnished several of the illustrative figures, for which acknowledgements in this handbook have been made.

The author expresses his appreciation to Colleen S. White for her assistance in typing, and to Edward R. Price for his editing, and the many facets of logistics involved in preparation of the manuscript, to William Horn for his drafting assistance and Gilbert Fitzpatrick for the cover design.

Other Books Published by ICT

1. Carstensen, Russell V., *EMI Control in Boats and Ships*, 1979.
2. Denny, Hugh W., *Grounding for Control of EMI*, 1983.
3. Duff, Dr. William G., *A Handbook on Mobile Communications*, 1980.
4. Duff, Dr. William G. and White, Donald R.J., Volume 5, *Electromagnetic Interference Prediction & Analysis Techniques*, 1972.
5. Feher, Dr. Kamilo, *Digital Modulation Techniques in an Interference Environment*, 1977.
6. Gabrielson, Bruce C., *The Aerospace Engineer's Handbook of Lightning Protection*, 1987.
7. Gard, Michael F., *Electromagnetic Interference Control in Medical Electronics*, 1979.
8. Georgopoulos, Dr. Chris J., *Fiber Optics and Optical Isolators*, 1982.
9. Georgopoulos, Dr. Chris J., *Interference Control in Cable and Device Interfaces*, 1987.
10. Ghose, Rabindra N., *EMP Environment and System Hardness Design*, 1983.
11. Hart, William C. and Malone, Edgar W., *Lightning and Lightning Protection*, 1979.
12. Herman, John R., *Electromagnetic Ambients and Man-Made Noise*, 1979.
13. Hill, James S. and White, Donald R.J., Volume 6, *Electromagnetic Interference Specifications, Standards & Regulations*, 1975.
14. Jansky, Donald M., *Spectrum Management Techniques*, 1977.
15. Mardiguian, Michel, *Interference Control in Computers and Microprocessor-Based Equipment*, 1984.
16. Mardiguian, Michel, *Electrostatic Discharge—Understand, Simulate and Fix ESD Problems*, 1985.

Other Books Published by ICT

17. Mardiguian, Michel, *How to Control Electrical Noise*, 1983.
18. Smith, Albert A., *Coupling of External Electromagnetic Fields to Transmission Lines*, 1986.
19. White, Donald R.J., *A Handbook on Electromagnetic Shielding Materials and Performance*, 1980.
20. White, Donald R.J., *Electrical Filters—Synthesis, Design & Applications*, 1980.
21. White, Donald R.J., *EMI Control in the Design of Printed Circuit Boards and Backplanes*, 1982. (Also available in French.)
22. White, Donald R.J. and Mardiguian, Michel, *EMI Control Methodology & Procedures*, 1985.
23. White, Donald R.J., Volume 1, *Electrical Noise and EMI Specifications*, 1971.
24. White, Donald R.J., Volume 2, *Electromagnetic Interference Test Methods and Procedures*, 1980.
25. White, Donald, R.J., Volume 3, *Electromagnetic Interference Control Methods & Techniques*, 1973.
26. White, Donald R.J., Volume 4, *Electromagnetic Interference Test Instrumentation Systems*, 1980.
27. Duff, William G., and White, Donald R.J., Volume 5, *Prediction and Analysis Techniques*, 1970.
28. White, Donald R.J., Volume 6, *EMI Specifications, Standards and Regulations*, 1973.
29. White, Donald R.J., *Shielding Design Methodology and Procedures*, 1986.
30. *EMC Technology 1982 Anthology*
31. *EMC EXPO Records 1986, 1987, 1988*

Notice

All of the books listed above are available for purchase from Don White Consultants, Inc., State Route 625, P.O. Box D, Gainesville, Virginia 22065 USA. Telephone: (703) 347-0030; Telex: 89-9165 DWCI GAIV.

PREFACE

Electrical *noise* has been a nemesis to the electronic industry for a long time, but giant strides have been made in recent years to overcome this unique phenomenon which is often referred to as Electromagnetic Interference (EMI) or more commonly called Radio Frequency Interference (RFI). Electrical *noise* has been an ever-increasing problem caused by the proliferation of all sorts of electrical and electronic devices and equipment. It has even reached such significant proportions that national and international regulatory commissions and societies have established standards to control the cause and effects of Electromagnetic Interference. Therefore, the burden falls on the design and test engineers, manufacturers, technicians, maintenance personnel, amateur radio operators and hobbiest and many, many others who are engaged in electrical or electronic production.

This handbook addresses the subject of *How to Control Electrical Noise* in basic uncomplicated terms with little emphasis on mathematics.

However, it is full of valuable information which would be useful to anyone in the field of electronics. It is highly illustrative and covers many subjects including the principal sources of *radio noise*, an ideal equipment box, grounding, shielding, cabling, bonding, filters, packaging, materials, trouble-shooting and numerous other important segments of information.

Your thoughts and impressions of the content and value of this handbook are solicited. If your reactions are favorable, you will certainly enjoy and benefit from our regular scheduled three-day intensified training courses on *Introduction to EMI/RFI/EMC*.

TABLE OF CONTENTS

HOW TO CONTROL ELECTRICAL NOISE

CHAPTER 1 ELECTROMAGNETIC INTERFERENCE CONTROL

CHAPTER 2 OUTSIDE SOURCES OF EMI AND THEIR REMEDIES

CHAPTER 3 THE FIRST BARRIER AGAINST RADIATED EMI

CHAPTER 4 HOW TO MAINTAIN BOX SHIELDING WITH COVERS

CHAPTER 5 WIRING

LIST OF ILLUSTRATIONS AND TABLES

CHAPTER 1   ELECTROMAGNETIC INTERFERENCE CONTROL

CHAPTER 2   OUTSIDE SOURCES OF EMI AND THIER REMEDIES

CHAPTER 5   WIRING

CHAPTER 6   BONDING

CHAPTER 7 GROUNDING SCHEME

CHAPTER 8 EMI FILTERS

CHAPTER 9 HINTS WHEN TROUBLESHOOTING ELEMENTARY EMI

TABLES

CHAPTER 1

ELECTROMAGNETIC INTERFERENCE CONTROL

The purpose of this book is to address the subject of Electromagnetic Interference (EMI) in uncomplicated terms, in the context of its definition, its effects, and to describe and illustrate practical and proven solutions on how to overcome common problems generated by the adverse effects of EMI. The phenomenon of EMI is often referred to as Radio Frequency Interference (RFI), Radio Noise, or just plain static.

Electromagnetics, having both magnetic and electrical properties, was at one time a mysterious phenomenon, and it still is to a much lesser extent. However, man has learned to understand, harness, and control electromagnetics to perform all sorts of incredible electro-mechanical functions. Without such knowledge, it would have been impossible to develop the basic electric generator and thus electric lights, let alone all of the other electronic devices and equipment that significantly contribute to the necessities and luxuries of every day life.

The growth and expansion of the electronics
industry have practically saturated the world
with electrical devices which generate and emit
electromagnetic energy from radiation, radio frequency waves, heat and light waves, X-rays et
cetera. Many emitters can cause spurious or unwanted signals which either directly or indirectly
interfere with the operation of other electronic
receivers or systems. Also physical phenomenon,
unrelated to man-made noise, such as lightning
can cause destructive interference and, then,
there are numerous instances and situations where
low-order interference becomes a common nuisance,
such as radio static, poor television reception
caused by CB transceivers, generator/motor start-
up, arc welders, auto ignitions, et-cetera. The
following describes many sources of such interference and corresponding remedies to related
problems.

1.1 The Threat From Electromagnetics

Whenever an electronic device or a piece of
electrical equipment creates electrical *noise*
that interferes with the performance of other
electrical equipment, or whenever such devices
are adversely effected by an external emission
noise source, then the problem of *ELECTROMAGNETIC*

INTERFERENCE is present. Examples of EMI are:
scratchy noises on Hi-Fi speakers whenever an
electrical switch is opened and closed, or when
a vacuum cleaner is operating; local CBs or amateur transceivers which interfere with TV sets;
telephones which are disrupted by transients
caused during lightning. Also computers can be
severely affected and caused to make errors by
static discharges between people and equipment,
particularly during dry seasons. Other significant examples are when industrial remote-control
devices are inadvertantly, or accidentally, trig-
gered by an electric welding machine, or when the
auto-pilot of an aircraft is jammed by a ground
based transmitter.

The ever increasing number of devices using
compact sensitive, or high-speed electronic components, make the incidents of interference more
frequent, particularly as new circuits become
more miniaturized and crowded into even less
space. All such EMI phenomena can be clearly
understood by a simple concept and/or properties
that they all share, i.e., the *source to victim*
concept (see Fig. 1.1).

When there is an EMI problem, there is always a *source* of the noise. The victim is the
device where the trouble or problem is occurring.

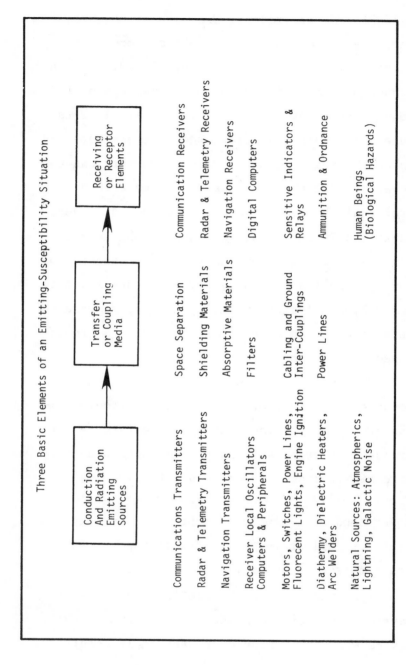

Figure 1.1 — Three Basic Elements of an Emitting/Susceptibility Situation.

1.4

In order for a noise source to cause interference there *MUST BE A COUPLING PATH* between the noise source and victim, *otherwise there would be no interference*. It is obvious, therefore, that the interference can be reduced, without entering sophisticated mathematics, at one or more of the following levels:

- The SOURCE - by decoupling, shielding, or simply making a noiseless design.
- The COUPLING PATH - by spacing or shielding if the coupling path is caused by radiation, or filtering if the coupling path is caused by conduction.
- The VICTIM - by local decoupling, isolating, shielding or by circuit redesign using less susceptible components.

Now, consider the numerical values involved (GETTING AWAY FROM THE DECIBEL's NIGHTMARE). To evaluate the amount of noise reduction provided by a *filter* or a *shield*, the expression is exactly like the one used by electricians for a voltage or current gain i.e.,

$$20 \log_{10}\left(\frac{V_{out}}{V_{in}}\right)$$

except, in this case, noise reduction is of

primary interest (not noise gain). For instance
in Fig. 1.2, the spike created by the refriger-
ator door-switch arcing is reduced by a small
filter on the power input of the Hi-Fi tuner (to
use the example of our preamble), which has a
ratio of 200 volts/2volts = 100 or 40 dB
attenuation. Table 1.1 gives the significance
of decibel as a ratio.

Similarly, a shield which attenuates the unwanted field by a factor of 10 is a barrier with
20 dB shielding effectiveness. In all cases, this
is a ratio of Volts/Volts, i.e., a *dimensionless
number*.

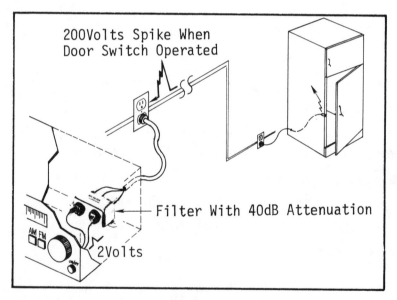

Figure 1.2 - Illustration of the Attenuation of
a Filter.

Table 1.1 - The Decibel as a Ratio

Voltage Ratio = $20 \log_{10}\left(\dfrac{V_1}{V_2}\right)$			Power Ratio = $10 \log_{10}\left(\dfrac{P_1}{P_2}\right)$		
dB	POWER RATIO	VOLTAGE RATIO	dB	POWER RATIO	VOLTAGE RATIO
0	1.00	1.00	10	10	3.16
1	1.26	1.12	20	100	10
3	2.00	1.41	30	1k	31.6
5	3.16	1.78	60	1M	1000
6	$\approx$ 4.00	2.00	-10	.1	.316
7	5.00	2.24	-20	.01	.10
9	$\approx$ 8.00	2.82	-30	.001	.0316
10	10.00	3.16	-60	10^{-6}	.001

As a general rule, the attenuation effectiveness in terms of dB values can be grouped as follows:

- 0 to 10 dB = Poor attenuation,
  A filter which reduces the conducted noise (or a shield reducing
  the EMI field) by this amount
  hardly pays for itself. The effect
  may be noticeable, but it cannot
  be relied upon to eliminate EMI.

- 10 to 30 dB = minimum range for
  achieving meaningful attenuation.

In mild cases, EMI would be eliminated.

- 30 to 60 dB = range where the average EMI problems can be solved.

- Over 60 dB = range for gaining above average attenuation - requires special attention and quality in shield and/or filter mountings (surface preparation, gasketing, bonding, etc.). Reserved for equipment which must operate at, or near, a 100% dependability factor and in extreme environments.

In expressing the amplitude of a noise voltage (when EMI is conducted), or the electromagnetic field which is creating noise, EMI specialists seldom use the terms Volt, or Volt/meter. Instead, they use their equivalent values in decibels, which means that a noise voltage of 10 volts is expressed as: $20 \log_{10}(10)$ decibels above 1 Volt, or 20 dB Volt, since $\log_{10}$ of 10 = 1.

The very low noise levels involved in EMI require the use of a smaller unit than the volt, i.e., the microvolt, or μV. Similarly, a noise amplitude of 100 μV, for instance, is expressed

as $20 \log_{10}$ (100 µV), i.e., $20 \log_{10}(100)$ decibels above one µVolt. Therefore, 100 µV is
equivalent to 40 dBµV. An electromagnetic field
strength of 1mV/meter, i.e., 1000 µVolts/meter,
is expressed as 60 dBµV/m. This is *no longer a*
dimensionless number, since x decibels above a
microvolt have the dimension of a voltage.

Table 1.2 gives a simple conversion between
µVolts and dBµVolts, or Volts and dB Volts.

Table 1.2 - Conversion of Volts (or Volts/m) and µVolts (or µVolts/m) in dBV and dBµV

Noise in Volts or Field in Volt/meter	Equivalent in µV or µV/meter	Corresponding dBVolts (or dBV/meter)	Corresponding dBµVolts (or dBµV/meter)
1000	10^9	60	180
316		50	170
100	10^8	40	160
32		30	150
10	10^7	20	140
3		10	130
1	10^6	0	120
.3	300,000	−10	110
.1	100,000	−20	100
.03	30,000	−30	90
.01	10,000	−40	80
.003	3,000	−50	70
.001	1,000	−60	60

CHAPTER 2

OUTSIDE SOURCES OF EMI AND THEIR REMEDIES

This section addresses the principal causes
of EMI (*noise*) which originate from sources external to the equipment, and the basic principles
which apply to protect the circuits from external
EMI. In these cases, it is assumed that no
action is possible to control the noise source
itself.

2.1 CW Transmitters

The atmosphere is entirely saturated by
electromagnetic fields generated by authorized
transmitters, which operate in a range from a
few tens of kHz to several 1000 MHz. To give an
idea, one could say:

- below .01 volt/meter (10 millivolts/
  meter) there is no risk of EMI (unless
  the product itself is an RF receiver
  tuned in the same frequency range).

- .1 volt/meter to 3 volts/meter is the
  beginning of potential EMI trouble
  (depending upon what frequency, and
  the physical dimensions of the equipment. The larger they are, the higher
  the risk).

- Above 3 volts/meter is a region of
  significant EMI risks, if no precau-
  tion is taken.

The basic protection strategy, therefore, is to
reduce the field strength locally received by
the equipment. This can be done by:

- using equipment casing to provide
  shielding,
- shielding cables feeding in or out of
  the equipment,
- using internal shields around the most
  susceptible parts of the unit,
- modifying the physical location or
  orientation of the equipment,
- using aluminum paper and screen-wire
  to shield the entire room where the
  the unit is installed.

If the culprit transmitter has been clearly
identified, as well as the portion of the victim's
circuitry which is disturbed, sometimes it is
simpler to design a stop-band *(trap)* filter to
eliminate a particular frequency.

2.2 High Frequency (Other-Than-Radio) Generators

Arc welders, ultrasonic machines, HF and
microwave ovens are well known outside EMI sources,
but there are more troublesome EMI sources, such

as computers, HF switching power converters, TV games or garage-door openers. Like the previous EMI sources discussed, they create both radiated and conducted noise which couple into equipment. In addition, because of their pulsed nature, they occupy a wide frequency bandwidth which creates a potential threat to a larger number of circuits (see Fig. 2.1). The solutions here are basically the same as those protecting against radio transmitters, except that these sources may be closer and difficult to locate, thereby requiring greater protection, such as:

- shielding of case and cables,
- physical relocation and/or reorientation,
- room shielding,
- filtering the power line, if this is a source of noise.

2.3 Broadband/Transient Sources

There is an infinity of random, non-intentional, noise sources which can surround equipment (see Fig. 2.2): They include:

- dimmer switches,
- burner igniters,
- dc and ac commutator motors,
- fluorescent lights, neon signs

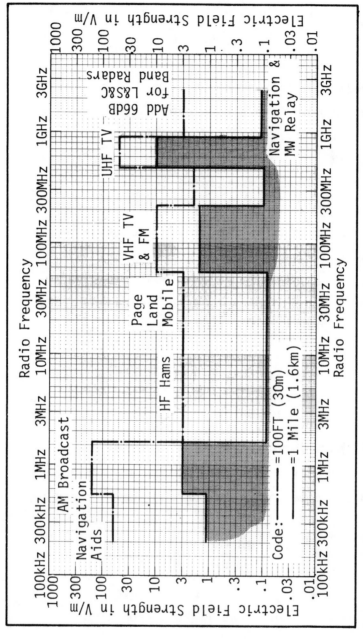

Figure 2.1 - Average Values of Electromagnetic Fields in Volt/meter caused by Radio Transmitters Located at 100 ft and 1 mile away.

Figure 2.2 - Examples of few among the many Broad-band Sources of Non-Intentional HF Energy which surround an Electronic Equipment.

• automobile ignitions

• HV overhead lines

The basic protection principles are the same as
in those shown in paragraphs 2.1 and 2.3.

2.4 Lightning

A lightning stroke creates a huge electromagnetic field and induces surge voltages in nearby power and communication lines (power, telephone,
video, etc.).   The basic protection methods include:

   • divert energy away with lightning
     rods and earthing conductors,

   • shielding, if lightning strokes are
     frequent and nearby,

   • installing transient protectors on in-
     put/output lines to prevent conducted
     surges from reaching equipment,

   • suppressing residual energy *inside*
     the equipment by installing secondary
     protection such as zener diodes or
     varistors (see Fig. 2.3).

2.5 Electro-static Discharge (ESD)

Electro-static discharge creates an enormous
number of problems (malfunctions or permanent
damages) in electronic circuitry which are troublesome to the engineer.   The basics of *what happens*
are shown in Fig. 2.4 when ESD involving a human

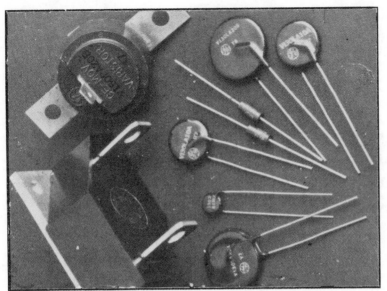

Figure 2.3 - Commercial Type of Lightning Transient Protection. *Courtesy of G.E.*

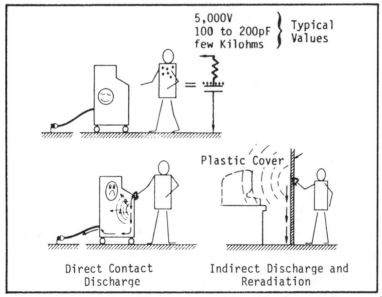

Figure 2.4 - The Electro-Static Discharge Scenario.

body occurs. This can also occur with any electrically charged body such as:

- plastic or fiber glass cans,
- paper
- rubber belts, et cetera.

Static discharges are aggravated by dry atmosphere, high personnel activity, nylon or wool carpeting etc. Such discharges can be reduced or eliminated by:

- controlling the relative humidity
- use of anti-static spray,
- using an ionizing air-gun,
- metallic casings or conductive paints
- shielding cables and ferrite beads on susceptible wiring.

2.6 Power Line Disturbances

Power line disturbances range from slow over-or-under voltage to sharp, extremely narrow transients, as shown in Fig. 2.5. The sources of such disturbances are often power-contact operations, heavy loads turning on/off, power semi-conductors operation, circuit breakers, or fuses blowing, lightning induced surges, etc. A good protection scheme should consider:

- power-line filters,
- dc distribution filtering,

- packaging of filter and power components.

- shielded power cords.

- relocation of equipment on another branch circuit.

- isolation (shielded) transformers.

- checking of grounding conductors at noise source.

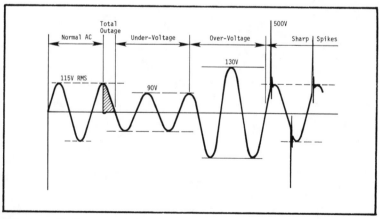

Figure 2.5 - Power Line Disturbances.

CHAPTER 3

THE FIRST BARRIER AGAINST RADIATED EMI

An *ideal enclosure* should act as a *continuously closed conductive envelope* in order to prevent *outside* fields from penetrating equipment and to prevent internally generated noises from escaping the enclosure. Such an envelope also protects from ESD and provides a conductive media to bond the internal cable screens or component shields (Fig. 3.1). Unfortunately, an *ideal box* is never achieved, because of ventila-

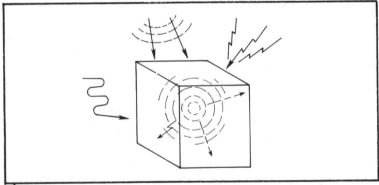

Figure 3.1 - Continuously Closed Conductive Envelope.

tion openings, maintenanace panels and doors, cover seams, cable-through holes, and connectors. However, for practical considerations in designing a box, note the following:

- try to keep the number and sizes of *openings* (sides, top, bottom) to the *minimum* compatible with their functions. (Figs. 3.2 and 3.3)
- cover ventilation openings or slots with *perforated grids* (1mm thick, 2-mm diameter holes).
- insure continuous electrical contact between grids and chassis (welding, or fastener spacing 10 cm)

The equipment box can be the first, and sometimes the only, barrier against radiation EMI problems. Think of the surface of a box as if it were a continuously conductive envelope. Every discontinuity, due to a door seam, screw or hinged-panel joint and/or assembled members, must have good electrical bonding qualities. Table 3.1 summarizes the surface treatments that may be considered in achieving good surface bonding, depending upon the severity of the EMI problem.

Figure 3.2 - Too Large, Unprotected Cable Entry.

Figure 3.3 - Cable Entry Blinded and Cable Shields Grounded at Box Entry.

Table 3.1 - Surface Treatment
(Depends also on type of Covers)

Simplest
and
Cheapest A) Painted Frame ⟶ Covers bonded by straps

B) Painted Frame + ⟶ Covers with finger stocks on
Riveted Strips corresponding surfaces (10 to
(tin coating over 50% of perimeter).
nickel plating).

C) Frame with conduc- ⟶ Covers with RFI seal or covers
tive paint (Technit with carbon gasket + magnetic
Silver-acrylic, strip (good pressure, good for
code 7300025 or acoustics).
ACHESON *Electrodag*
code 405 or conduc-
tive, corrosion-
free plating.

D) Entirely Plated ⟶ Cover with 100% finger stocks.
Frame (Tin coating
over Nickel Plating)

Most Effective
...and costly

3.4

Solution A only requires braided jumpers (Fig. 3.4). Even though only one would seem to be sufficient to electrically ground the panel to the frame, it is recommended that several jumpers be used and the spacings between them never exceed $\lambda/10$, where λ is the wavelength corresponding to the frequency where the potential EMI occurs. Remember that wavelength and frequency are related by:

$$\lambda_{meters} = \frac{300,000}{F_{kHz}}, \quad \text{therefore,}$$

When:

F = 1 kHz	λ = 300,000 meters
F = 1 MHz	λ = 300 meters
F = 30 MHz	λ = 10 meters
F = 100 MHz	λ = 3 meters
F = 300 MHz	λ = 1 meter

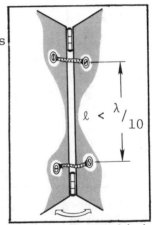

Figure 3.4 - Solution A, Braided Jumpers Screwed on Paint Free Areas.

If the EMI frequency is unknown, use grounding straps spaced not more than 20 cm (8" apart). Do not place straps near sensitive items.

Solution B is an acceptable compromise which insures some continuity, not only on the hinged side, but on the entire perimeter of a panel.

Several types of finger stocks are available in two categories; *the low-pressure, knife-edge,* and the *medium-pressure spring contact* (Par. 4.3). Both types require adequate retention, proper flattening control and hinge adjustment.

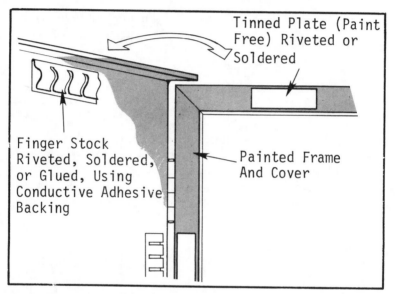

Figure 3.5 - Solution B - To Maintain Shield Continuity with Painted Metal Box.

Solution C relies on a conductive coating over two mating surfaces. The coating acts as a paint to protect the metal against corrosion, and insures a good conductive media between each metal surface and the gasket. It also has the advantage of being effective in the case of plastic boxes.

Solution D is the most efficient since 100% of
the seam becomes a very good conductive joint
(Fig. 3.6). Besides its cost, it adds the need
for a strong locking mechanism to insure good
even pressure on all of the spring blades. This
method is applicable to both rotating (hinged)
or slide-mating surfaces.

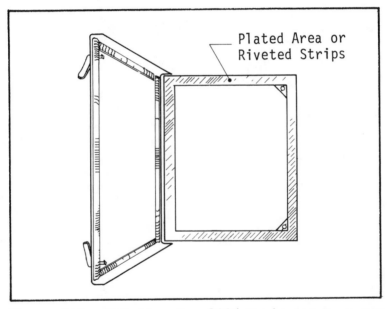

Figure 3.6 - Solution D - 100% Perimeter Coverage
By Fingerstocks.

CHAPTER 4

HOW TO MAINTAIN BOX SHIELDING WITH COVERS

4.1 Cover Material

A metal cover is by far the best shielding
that can be used. Steel is preferable to aluminum (or copper) due to good absorption loss provided its thickness is > 1mm. Below that thickness, aluminum or copper make equivalent or
better shields, especially against close magnetic
fields, because of their superior conductivity
which gives good reflection losses for low frequencies where absorption is moderate.

Plastic covers provide no shielding whatso-
ever. Therefore, to achieve some attenuation the
following must be used:

- conductive plastic (conductive spray
  or mass inclusion of conductive
  particles) or,
- internal shields over susceptible (PC
  boards, cable fanout) or radiating
  (power supplies, CRT) areas. In extreme cases, special shields are needed:
  Nickel alloyed steel or Mu metal (its
  magnetic permeability μ_r is 60 to 80
  times better than ordinary steel, up
  to about 100 kHz).

4.2 Cover Packaging

Provide overlapping of cover edges, even if there is no EMI gasket (see Fig. 4.1). <u>DO NOT PLACE TOO CLOSE TO COVERS</u>, and especially cover seams:

- Sensitive items = low-level logic, analog circuits, and high-speed logic.
- EMI sources = oscillators, transformers, relays, coils, and high-speed logic cards.

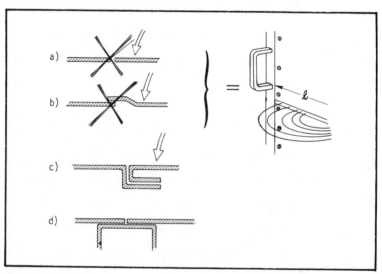

Figure 4.1 - a) and b) shows edge mating of covers acting as a radiating slot. However good the shield effect was, it is merely reduced to the ratio of ℓ to $\lambda/2$, the half-wave length of the EMI field. In c) and d) the overlaps act as a capacitive coupling between the two covers and provides continuity.

4.3 Covers-to-main Frame Bonding (see also box shielding in Chapter 3)

Maintain a good electrical continuity between covers and main frame (see Fig. 4.2)

- with *STRAPS* – place strap far from sensitive items,
- with *FINGERSTOCKS* – partially or 100% require adequate retention, flattening (5 to 1.5mm) and hinge adjustment,

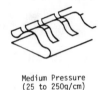

Low Pressure

Medium Pressure
(25 to 250g/cm)

- with *EMI GASKET* + conductive paint, or plating (beware of gasket flattening).

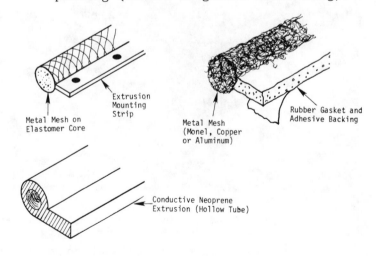

Metal Mesh on
Elastomer Core

Extrusion
Mounting
Strip

Metal Mesh
(Monel, Copper
or Aluminum)

Rubber Gasket and
Adhesive Backing

Conductive Neoprene
Extrusion (Hollow Tube)

4.3

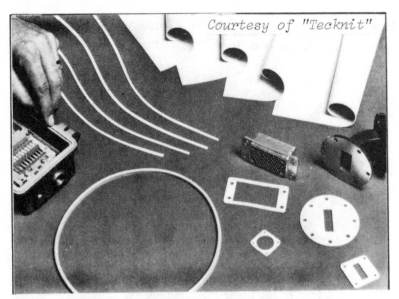

Typical Conductive Elastomer Gaskets

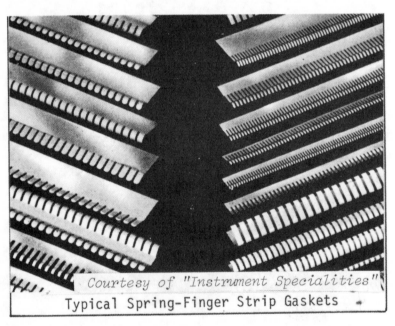

Typical Spring-Finger Strip Gaskets

Figure 4.2 - Elastomer and Spring-Finger Gaskets.

CHAPTER 5
WIRING

To avoid inter-cable noise coupling (cross-talk) during the early design stage, identify and categorize the separate cable routings.

For example:

Category (1) ac power cables ⟶ Noise carriers

Category (2)* dc distribution ⟶ Noise carriers and/or victims

Category (3) signal and logic cables ⟶ Victims

$\begin{cases} (3a) & \text{Analog, low-level signals} \\ (3b) & \text{digital signals} \end{cases}$

* Signal wires from relays, circuit breakers, thermal switches, etc, fall in category (2).

5.1 Routing

- Category (1) - route along frame members and bottom of the machine/equipment,
- Category (2) - route along frame members and sheet metal plates, but separate from (1) (avoid *open space hanging*),
- Category (3) - route far from (1) and (2).

As a general rule, cables from subcategory (3b), carrying digital signals, should be spaced 2.5cm (1") from Category (1) for every 1 meter of their possible parallel run. For cables from subcategory (3a), (depending upon the sensitivity of the

circuits with which they interconnect), a spacing of 25 cm (10 inches) from category (1) for
every meter is required to protect an analog
circuit having 10 mV sensitivity. Remember that
power wiring not only carries 60 Hz voltages, but
it also has associated line spikes which couple
strongly to nearby victim wiring. For example:

    If a packaging constraint forces
    analog signal cables to be run par-
    allel to ac cables over a distance
    of 0.3 meter (1 ft.), a separation
    of 25 cm x 0.3 = 7.5 cm (3 inches)
    should be provided.

Digital cabling (3b) can in turn be a noise car-
rier for more susceptible wiring like (3a).
However, considering the low levels involved, it
is a sufficient practice not to lace them together
in the same harness. In all cases, at intersec-
tions, make right angle runs.

At machine entry holes, use separate loca-
tions for categories (1), (2) and (3) points of
entry. Cable routings outside the equipment
should also be controlled to avoid category mixes,
whenever possible (see Fig. 5.1), and re-radiation
(Fig. 5.2).

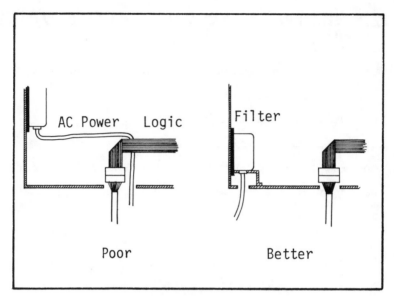

Figure 5.1 - Cable Entries.

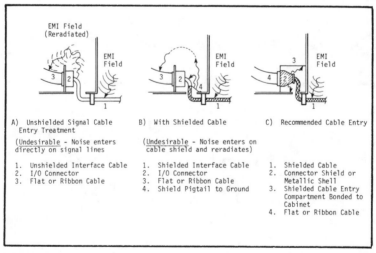

Figure 5.2 - *Do's* and *Don'ts* for Cable Entry at Box Interface.

5.2 Shielding of AC Cables - Category (1)

Shielding of an ac cable is not mandatory.
It depends upon equipment operation technology,
noise susceptibility of logic, speed of logic,
type of power supply and power mains ambient
noise. If an efficient filter is required, or
good low-impedance ground is necessary, shielding
the power cord is a good and effective supplemen-
tal practice.

For Shield Termination of Power Cord (see also
filter installation):

Machine side: Extend shield up to the line
filter, which should be *closest* to the power cord
entry hole and main $\perp$. Connect the shield to
the chassis ground (preferred to a jumper) with
a 360° metal clamp (Fig. 5.3).

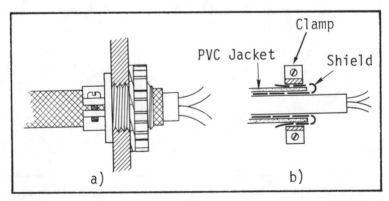

Figure 5.3 - Correct Way to Terminate a Shielded
Cable at Box Entry.

Plug side: If there is a good (or dedicated)
earth wire, connect the shield to the ground pin
along with the green wire (Fig. 5.4) or, better
still, connect it with a 360° clamp to a metallic
plug cap, when available. If a very noisy ground
is suspected, it is better not to connect the
shield to ground pin, plug side. Using a circu-
lar clamp, if the PVC jacket is to be fully in-
serted into the clamp, or strain relief, this
can be accommodated by folding the shield back
over the jacket (b). Grip type bushings like
used in electrical wiring (Thomas & B., etc) are
excellent for shield termination.

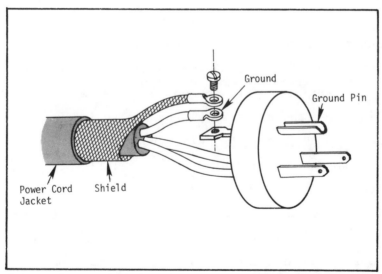

Figure 5.4 - Terminating a Power Cord Shield on
the Plug Side, When you can Depend on a Good
Earthing Wire.

5.3 EMI Protection of DC Distribution Category (2)

- Route every hot wire with its 0v return (no loop surface),
- Use capacitor decoupling on board pins (differential mode) or feed-thru dc capacitors (common-mode),
- On power printed circuit, keep copper planes as large as possible (decreases ripple and HF noise, also cheaper (see Fig. 5.5).

For PC cards which are not purely Power Supply cards but a mix of bulky, large current and low-level electronics, it is good practice *to leave* copper on all areas which have not been used for traces. Such planes can be on the component side (Fig. 5.6a) and connected to the ground bus, or on the non-component side. In this case, the plane should be broken into a ground grid (Fig. 5.6b) with about 50% voids to preclude flow-soldering problems. Table 5.1 gives the value of printed wiring impedance (combined resistance + inductance) with frequency increase. Note that no printed trace other than a copper plane can achieve a lower HF impedance.

For dc distribution on cards or motherboards with a high density wiring, where

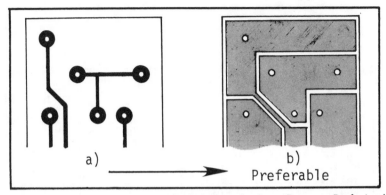

Figure 5.5 - Recommended Layout for Power Printed Circuit.

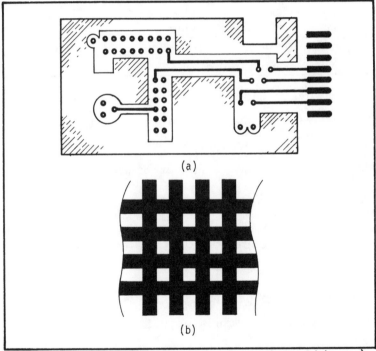

Figure 5.6 - Copper Plane on Component Side a) and on non-component side b). Grid is to avoid Wave Soldering Problems without compromising the Low-impedance Plane Concept.

Table 5.1 - Impedance of Printed Circuits

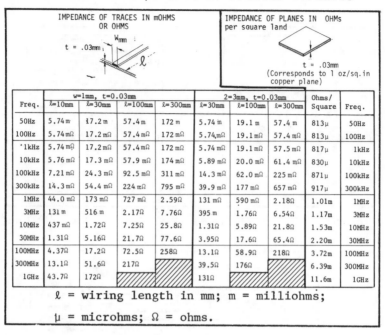

Freq.	w=1mm, t=0.03mm				2=3mm, t=0.03mm			Ohms/ Square	Freq.
	ℓ=10mm	ℓ=30mm	ℓ=100mm	ℓ=300mm	ℓ=30mm	ℓ=100mm	ℓ=300mm		
50Hz	5.74m	17.2m	57.4m	172m	5.74m	19.1m	57.4m	813µ	50Hz
100Hz	5.74mΩ	17.2mΩ	57.4mΩ	172mΩ	5.74mΩ	19.1mΩ	57.4mΩ	813µ	100Hz
1kHz	5.74mΩ	17.2mΩ	57.4mΩ	172mΩ	5.74mΩ	19.1mΩ	57.5mΩ	817µ	1kHz
10kHz	5.76mΩ	17.3mΩ	57.9mΩ	174mΩ	5.89mΩ	20.0mΩ	61.4mΩ	830µ	10kHz
100kHz	7.21mΩ	24.3mΩ	92.5mΩ	311mΩ	14.3mΩ	62.0mΩ	225mΩ	871µ	100kHz
300kHz	14.3mΩ	54.4mΩ	224mΩ	795mΩ	39.9mΩ	177mΩ	657mΩ	917µ	300kHz
1MHz	44.0mΩ	173mΩ	727mΩ	2.59Ω	131mΩ	590mΩ	2.18Ω	1.01m	1MHz
3MHz	131m	516m	2.17Ω	7.76Ω	395m	1.76Ω	6.54Ω	1.17m	3MHz
10MHz	437mΩ	1.72Ω	7.25Ω	25.8Ω	1.31Ω	5.89Ω	21.8Ω	1.53m	10MHz
30MHz	1.31Ω	5.16Ω	21.7Ω	77.6Ω	3.95Ω	17.6Ω	65.4Ω	2.20m	30MHz
100MHz	4.37Ω	17.2Ω	72.5Ω	258Ω	13.1Ω	58.9Ω	218Ω	3.72m	100MHz
300MHz	13.1Ω	51.6Ω	217Ω	▨	39.5Ω	176Ω	▨	6.39m	300MHz
1GHz	43.7Ω	172Ω	▨	▨	131Ω	▨	▨	11.6m	1GHz

ℓ = wiring length in mm; m = milliohms;

µ = microhms; Ω = ohms.

low-impedance planes cannot be achieved, use
flat buses for power distribution (Fig. 5.7).
They are better against EMI than solid wires,
and also decrease the number of discrete decoupling capacitors.

5.4 EMI Protection of Logic/Signal Cabling Category (3)

At this point, it will be useful to distin-
guish between the kinds of noises generated by
differential-mode and *common-mode*.

Figure 5.8 shows that differential current
I_{DM} is the *normal* useful (symetric) current flowing to and from the load. A corresponding differential voltage can be measured wire-to-wire.
The common-mode current I_{cm} (asymmetric) flows

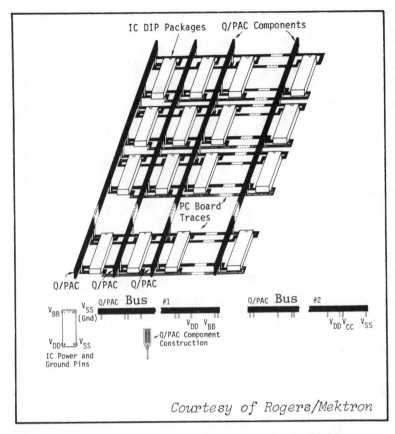

Figure 5.7 - Power Distribution Network in a Matrix of 16k MOS Dynamic RAM Services. Each "Q Pac" bus bar is soldered to perpendicular V+ and 0_V printed traces so a low impedance "grid" is achieved even though there is *No Ground Plane.*

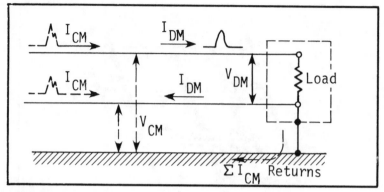

Figure 5.8 - Differential Mode and Common Mode
Noise.

in both wires in the *same direction* at the same
time. The corresponding common-mode voltage V_{CM}
cannot be measured wire-to-wire, but it can be
measured wire-to-ground.

To reduce EMI coupling from other cables,
and outside EMI ambients as well, in all cases,
each signal wire should have *its own independent
return* running closely (no loop area, no common
impedance return shared with other signals). For
the three principal kinds of wiring:

- *FLAT CABLE AND RIBBON*
  Each wire is surrounded by
  1 or 2 ground wires ≃ equivalent
  to a shield.

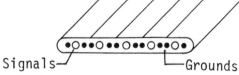

- *TWISTED PAIR*

  Each wire runs with
  its return wire: very
  efficient against differential
  mode noise; inefficient for common mode noise.

- *COAX*

  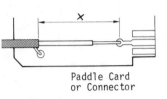

  Paddle Card
  or Connector

  - The shield serves as
    0v common return (Low
    Z). Connect the shield
    to 0v dc at *both ends*.

  - Stripped portion (X) may
    ruin shield effect if too long.

  - Disadvantage: the 0v surrounding the center
    wire may be disturbed by radiated or conducted
    noise.

- *SHIELDED TWISTED PAIR*

  Very efficient against $\begin{cases} \text{differential mode} \\ \text{noise (twisting)} \\ \text{common mode noise} \\ \text{(shield)} \end{cases}$

  For *low* frequency and high susceptibility, connect the shield to *Frame Ground* at one end
  only (receiving side unless there are other
  constraints).

  For *high* frequency, connect the shield to
  Frame at both ends (must be discussed with
  an EMC specialist).

In all cases, connect unused wire ends to 0V or
to a dummy load (never floating). Connect magnetic head cases and connector housings to Frame
Ground. In extreme cases (very high frequency
transients, electro-static discharge) install
ferrite suppressors around the cable (Fig. 5.9).

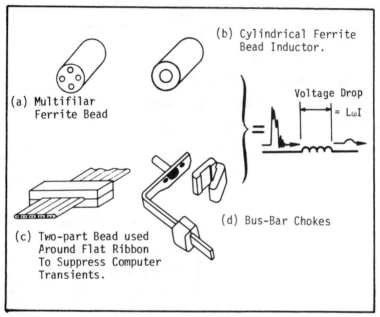

Figure 5.9 - Typical Common Mode Suppression
Ferrite.

Ferrites work both by the added series inductance and equivalent resistance and start to
be efficient above few MHz. They behave like an
added resistor which would *show up* for noise
only, with values ranging from 1 MHz to 100 MHz

of about 40Ω for (a) and (b) and for an outside
diameter of ≃ .3 cm and a length of ≃ .6 cm, and
about 150Ω for (c) and type (d) which is a special
version for large conductors.

5.5 Improving Cabling Immunity to EMI

In addition to the preceding suggestions,
one very classical way to improve cable immunity
(applied since the early times of telephony) is
to use balanced isolation transformers which are
available from the very low frequencies (audio
applications) up to 20 MHz or more (Fig. 5.11).
The basic improvement is shown in Fig. 5.10.
Without EMI care (a), the EMI noise developed between A and B creates a loop current which spoils
the normal transmitted signal, especially if the
transmission is unbalanced. Breaking the loop by
isolation transformers (b), prevents I_{cm} from circulating since the transformer (ideally) should
not process common-mode signals. In (c), the
driver and receiver are balanced (they talk with
+ and - signals referred to 0_{volt}). In (d), the
unbalanced driver and receiver are balanced by
using a BALUN (BALanced-UNbalanced) transformer.
The BALUN has a shield to reduce the primary-to-
secondary parasitic coupling, which has been
grounded. A twisted pair is recommended to avoid

5.13

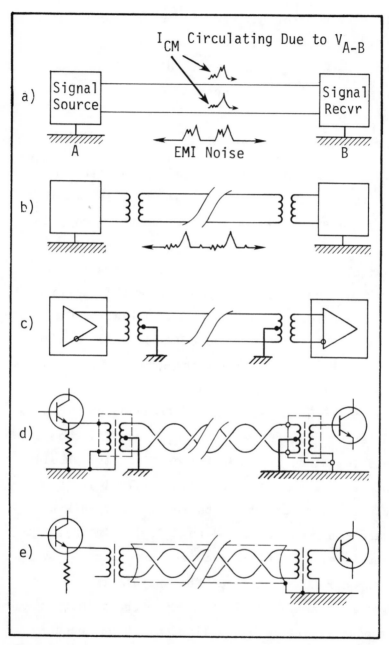

Figure 5.10 - Using Balanced Transformers.

field pick-up by the pair itself. Finally, in
(e) a total protection is achieved by adding a
shield, which is grounded to the receiver side
only (also to avoid circulating current in the
shield).

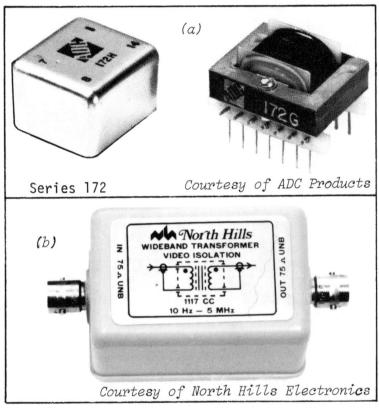

(a)

Series 172   *Courtesy of ADC Products*

(b)

Courtesy of North Hills Electronics

Figure 5.11 - Example of Commercial Types of Isolation Signal Transformers; Type a) is for audio
applications (.2 to 4kHz); Type b) is for fast
clocks and video applications (10Hz to 5MHz).
They provide common mode isolation from 40 dB to
60 dB, i.e., only 1% or .1% of the common mode
noise appears on the Output Terminals.

CHAPTER 6

BONDING

Bonding is involved in the solution of all problems encountered when a good electrical contact is relied upon for EMI purpose. Bonding is also required for safety. A good safety bond (DC resistance ≃ 0) may be poor for EMI protection if its impedance becomes large at high frequencies (Fig. 6.1)

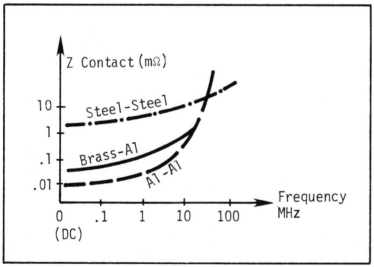

Figure 6.1 - Degradation of Bonding Impedance with Frequency Increase.

The major considerations are:

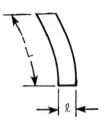

- *STRAPS*

  Prefer braid to jumpers.

  Thin, wide and shortest

  possible.

  Regarding the length, a strap should

  have theoretically $\ell \geq L$. In practice,

  try to achieve L not larger than 10 x ℓ.

  For straps which may be disconnected

  (servicing) use Slip-On contacts

  rather than screws.

- *CONTACT RESISTANCE*

  Increased by corrosion $\Big\{$ Galvanic

  $\qquad\qquad\qquad\qquad\qquad\qquad$ Electrolytic

Contacts between metals
of adjacent groups are
compatible. With dis-
similar metals, use a
third intermediate metal.
With Al, avoid star-
washer, use Nickel coat-
ing or tinned washers.
Coat metal interface
with sealant to stop
moisture penetration.

Magnesium & Alloys
Zinc
Aluminium & Alloys   + susceptible
Cadmium
Carbon steel, iron
Ni-Cr Stainless Steel
Lead, Tin
Nickel
Copper, Brass
Silver-Nickel   + protected
Chromium Steel   (immune)
Silver, Gold, Titanium, Graphite

● *AVOID INDIRECT BONDING*

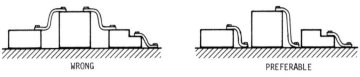

WRONG   PREFERABLE

● *AVOID MOTION OF CONTACT SURFACES*

CHAPTER 7

GROUNDING SCHEME

Regardless of the product or equipment being worked upon, whether it is a small home project or a large industrial system, it is mandatory that the grounding scheme be determined and under-stood. While grounding techniques may solve some EMI problems, ironically many EMI troubles occur because of uncontrolled grounding configurations which create deliberate or *ghost*-ground loops. There are two main types of grounding options; *centralized* (single-point) or *distributed* (multi-point). As long as wiring distances are less than $\lambda/10$ at EMI frequency (refer to page 3.5, Chapter 3) the *single-point (star) ground* (see Fig. 7.1) is the most achievable and compat-ible with other constraints. Beware of the dra-matic increase in impedance of regular wires when frequency and length increase (Table 7.1). The main rule is to eliminate any 0_{volt}-to-chassis loops, which are the most troublesome. It also

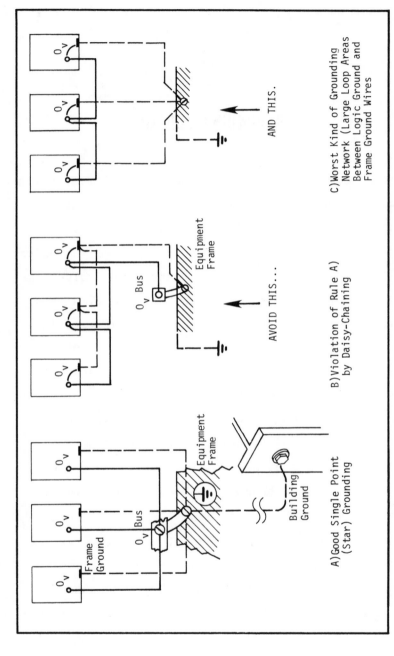

Figure 7.1 - Typical Centralized (Star) Grounding with some Typical Mistakes.

Table 7.1 - Impedance of Straight Circular Copper Wires

FREQ.	AWG#-2, D=6.54mm				AWG#-10, D=2.59mm				AWG#22, D=.64mm			
	ℓ=1cm	ℓ=10cm	ℓ=1m	ℓ=10m	ℓ=1cm	ℓ=10cm	ℓ=1m	ℓ=10m	ℓ=1cm	ℓ=10cm	ℓ=1m	ℓ=10m
10Hz	5.13μ	51.4μ	517μ	5.22m	32.7μ	327μ	3.28m	32.8m	529μ	5.29m	52.9m	529m
50Hz	5.20μ	55.5μ	624μ	7.16m	32.8μ	392μ	3.30m	33.2m	530μ	5.30m	53.0m	530m
100Hz	5.41μ	66.7μ	877μ	11.2m	32.9μ	332μ	3.38m	34.6m	530μ	5.30m	53.0m	530m
300Hz	7.32μ	137μ	2.19m	30.4m	33.7μ	365μ	4.11m	46.9m	530μ	5.30m	53.0m	531m
1kHz	18.1μ	429μ	7.14m	100m	42.2μ	632μ	8.91m	116m	531μ	5.34m	53.9m	545m
3kHz	52.5μ	1.28m	21.3m	300m	86.3μ	1.65m	25.0m	336m	545μ	5.71m	60.9m	656m
10kHz	174μ	4.26m	71.2m	1.00Ω	268μ	5.41m	82.9m	1.11Ω	681μ	8.89m	113m	1.39Ω
30kHz	523μ	12.8m	213m	3.00Ω	799μ	16.2m	248m	3.35Ω	1.39m	22.0m	305m	3.91Ω
100kHz	1.74m	42.6m	712m	10.0Ω	2.66m	54.0m	828m	11.1Ω	4.31m	71.6m	1.00Ω	12.9Ω
300kHz	5.23m	128m	2.13Ω	30.0Ω	7.98m	162m	2.48Ω	33.5Ω	12.8m	214m	3.01Ω	38.7Ω
1MHz	17.4m	426m	7.12Ω	100Ω	26.6m	540m	8.28Ω	111Ω	42.8m	714m	10.0Ω	129Ω
3MHz	52.3m	1.28Ω	21.3Ω	300Ω	79.8m	1.62Ω	24.8Ω	335Ω	128m	2.14Ω	30.1Ω	387Ω
10MHz	174m	4.26Ω	71.2Ω		266m	5.40Ω	82.8Ω		428m	7.14Ω	100Ω	1.29kΩ
30MHz	523m	12.8Ω	213Ω		798m	16.2Ω	248Ω		1.28Ω	21.4Ω	301Ω	3.87kΩ
100MHz	1.74Ω	42.6Ω			2.66Ω	54.0Ω			4.28Ω	71.4Ω	1.00kΩ	12.9kΩ
300MHz	5.23Ω	128Ω			7.98Ω	162Ω			12.8Ω	214Ω	3.01kΩ	38.7kΩ

* AWG = American Wire Gage
   D   = wire diameter in mm
   ℓ   = wire length in cm or m
   μ   = microhms
   m   = milliohms
   Ω   = ohms

   Non-Valid Region
   for which ℓ ≥ λ/4

avoids the ground return current of one building
block to flow thru another block return wire,
and possibly a build-up of some noisy voltage
drops. Therefore, each building block has its
0_V floated from the box, and the box itself hard-
wired to the main earth terminal of the machine/
equipment.

The *distributed (multipoint) ground* is to
be used when the wire length is in excess of $\lambda/4$
and prevents a dependable star ground. In this
case, every 0_{volt} is grounded locally to the near-
est chassis point. To perform well, the portion
of chassis between subassemblies (1) and (2) in
Fig. 7.2 should have the lowest impedance possible,
without discontinuities, in order to approach a
perfect conductive plane. If there are mechani-
cal discontinuities they should be made conductive as described in Chaps. 4, *Covers* and 6,
Bonding.

Even though both 0_V are tied to the chassis,
the latter should not be relied upon for active
signal return, if box #1 and #2 (Fig. 7.2) talk
together: never miss the dedicated return wire.

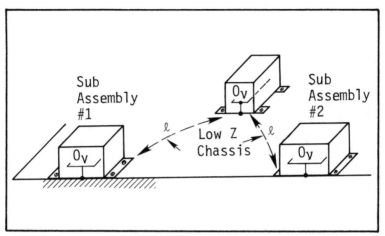

Figure 7.2 - Distributed (multipoint) Ground,
when a good low Impedance is available and EMI
Frequency is very high ($\ell \geq \lambda/4$).

CHAPTER 8

EMI FILTERS

8.1 Selecting The Right Filter

Filters must be carefully selected and the selection should not be based upon a vague belief in a manufacturer's catalog. A filter which has been good for one type of equipment may be inefficient for another type, or even too efficient. In this latter case, the filter may be more than what is needed in both cost and packaging aspect, providing for example a 1/1000 attenuation (60 dB) while the real need was to decrease the noise by a ratio of 1/50 (34 dB). Think of a filter as having a double role:

- first, it attenuates the *incoming* noise from other sources,

- second, it attenuates the noise *generated within the equipment itself*. This noise level is regulated by military specifications like MIL-STD-461B, or commercial specifications like FCC Parts 15 & 18 or VDE 871 & 875.

A proper selection of a power main filter should consider:

- Highest nominal voltage,

- Highest, RMS input current (corresponding to the lowest voltage input tap setting, and the maximum optional features installed),

- Type of power supply – switching power supplies require an efficient filtering at low frequencies because of their high noise generation,
- Nature of the electronic circuit inside of the equipment: low-level analog circuits, and fast logic (as victims) and logic with significant transition current (as noise source) all require a well filtered power supply,
- Permissible leakage current in the safety wire – National Electrical Code and Worldwide organizations (IEC) require this current to be less than:
  - 5mA for equipment with ground wire,
  - .5mA for class II equipment (double insulated).

Figure 8.1 is a reminder of what should be watched for in terms of filter data.

8.2 Low Cost Filters

A relatively inexpensive and rugged filter, made of a few discrete components, may be sufficient for less susceptible equipments, which

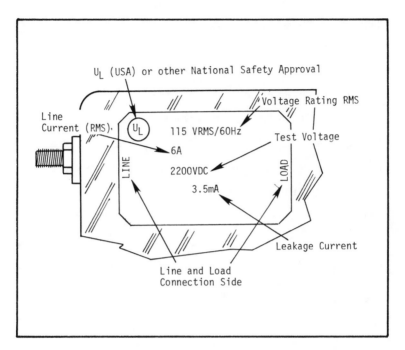

Figure 8.1 - Example of Essential Filter Data.

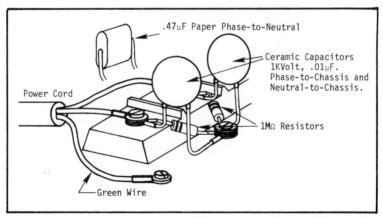

Figure 8.2 - *Home-Made* Filter for less Susceptible Machines.

have slow speed and good immunity circuits,
linear power-supply regulators with efficient output decoupling, or low-power devices. Also the
performance of purchased filters may be upgraded
by the addition of a capacitor (Fig. 8.2). In
the case of high power ac or dc, where large currents require an expensive choke, use feed-thru
capacitors (Fig. 8.3).

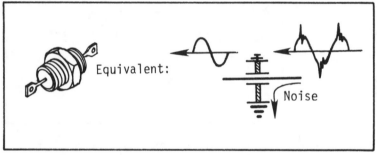

Figure 8.3 - Feed-through Capacitors.

Use High Frequency Caps (Ceramic,
Polystyrene, Paper).
Capacitor A traps
differential noise,
B traps common mode

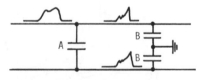

and beware of lead length (keep as short as
possible).

8.3 Filter Location

Always put filters as close as possible to
the power cord entry. Ideally, the filter should

be the first component the power cord encounters when it passes through the equipment frame. Always keep the filtered output wiring from the input wiring, and insure a perfect grounding of the filter case (paint-free, tinned areas; see Figs. 8.4 and 8.5.

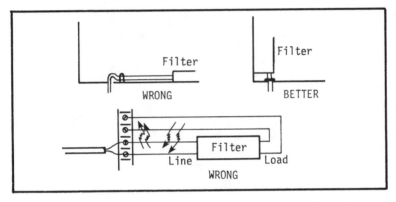

Figure 8.4 - *Do's and Don'ts* in Filter Mounting.

8.4 Transient Suppressors Other Than Filters

Filters do not actually dissipate unwanted energy by their mode of operation, rather they repel undesired signals back toward the source (ac mains or ground). Other types of components can clamp large over-voltages that regular filters cannot handle. They are *varistors* and *gas-tube arrestors* (Fig. 8.6). These components are especially recommended for protection against *lightning-induced surges*, not only from power

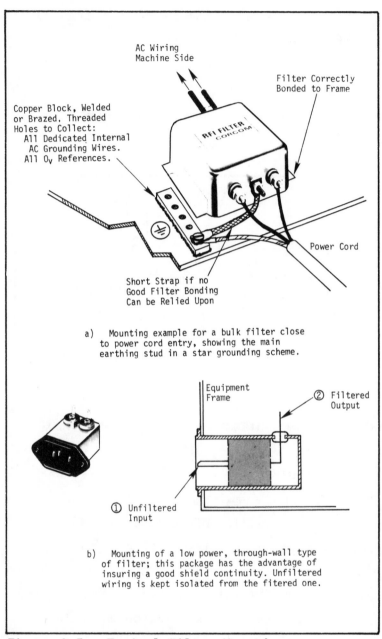

AC Wiring
Machine Side

Filter Correctly
Bonded to Frame

Copper Block, Welded
or Brazed. Threaded
Holes to Collect:
  All Dedicated Internal
  AC Grounding Wires.
  All 0_V References.

RFI FILTER
CORCOM

Power Cord

Short Strap if no
Good Filter Bonding
Can be Relied Upon

a) Mounting example for a bulk filter close
    to power cord entry, showing the main
    earthing stud in a star grounding scheme.

Equipment
Frame

② Filtered
  Output

① Unfiltered
  Input

b) Mounting of a low power, through-wall type
    of filter; this package has the advantage of
    insuring a good shield continuity. Unfiltered
    wiring is kept isolated from the fitered one.

Figure 8.5 - Typical Filter Mountings.

8.6

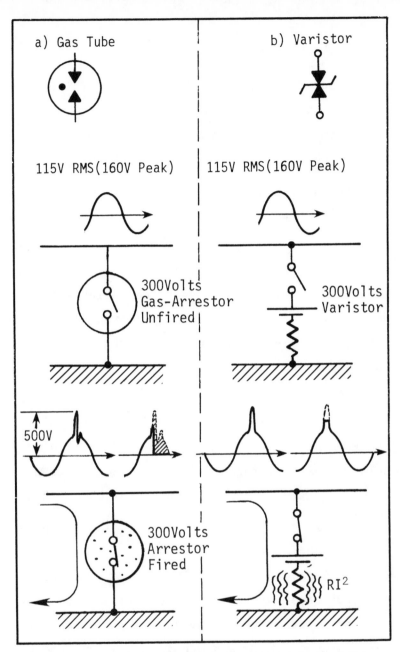

a) Gas Tube

b) Varistor

115V RMS(160V Peak)

115V RMS(160V Peak)

300Volts
Gas-Arrestor
Unfired

300Volts
Varistor

500V

300Volts
Arrestor
Fired

RI^2

Figure 8.6 - Basic Operation of 2 kinds of Transient Protectors: a) the gas tube practically shorts everything (spike and normal voltage) once its threshold voltage is reached; in b) the varistor *clamps* the voltage to a preset value.

lines, but from any outdoor line or serial exposed to frequent lightning strokes, i.e., telephone lines, television or radio antennas.

In Fig. 8.6a the gas tube practically shorts everything (spike and normal voltage) once its threshold voltage is reached.

As long as a surge voltage does not exceed the preset value, the gas tube is simply an open circuit. However, as soon as the firing threshold is reached, the gas tube performs as a closed switch, which short-circuits the surge. Once the surge is ended, the device stops to conduct and recovers its initial status.

In Fig. 8.6b the varistor *clamps* the voltage to a preset value. Varistors often require a low value resistance on the line before them to avoid destruction by too high surge currents.

Figure 8.7a shows a power or telephone protection scheme, line-to-line and line-to-ground (common mode), and Fig. 8.7b shows a schematic for protection of radio and television antennas. Another alternative, is to use only one arrester for the center wire and ground the outer conductor (shield) directly to all possible *grounded* metallic part of the roofing (water pipes, gutters, etc.). However, this may create a

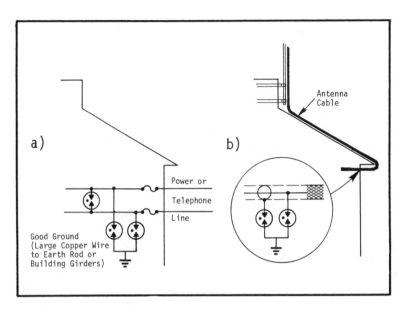

Figure 8.7 - Basic Installation of Surge
Arrestors.

permanent background noise. Rather than letting
unwanted noise pollute all the internal circuitry
until it reaches the filter or suppressors, it
is cost effective to eliminate it as close as
possible near the component where it is created.
Some sources of internal noise are shown in Fig.
8.8.

SOURCE	SUPPRESSION
Contactor, Relay	• Diode (Fast Recovery) • RC Circuit (keep leads ultra short) 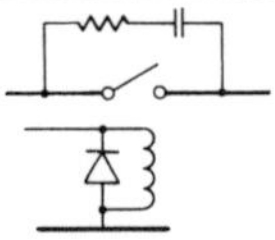
SCR's, Power Diodes 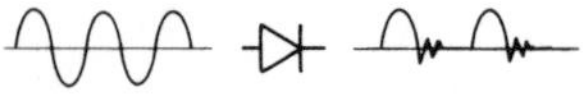	• Ferrite Core • RC Circuit in // • High Frequency Capacitors in // with Electrolytic Capacitors 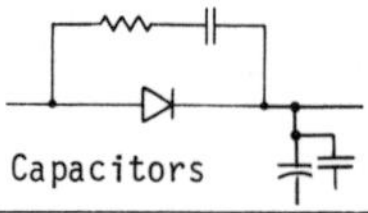
DC Motors (Commutators)	High Frequency Capacitors 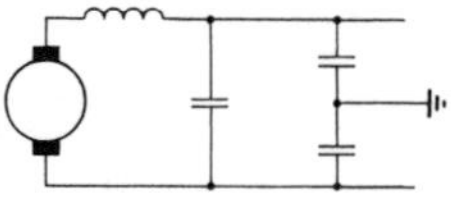
Fans (Especially those with the rotor outside)	Shield susceptible circuits or locate fan away from circuits
Transformers 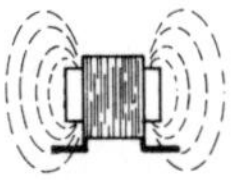	Change angular position between transformer and wires (wiring must not cross any lines of flux)

Figure 8.8 - Some Sources of Electrical Noise (Internal).

CHAPTER 9

HINTS WHEN TROUBLESHOOTING ELEMENTARY EMI

Real EMI measurements require sophisticated and sensitive equipment like spectrum analyzers, EMI receiver with accurate bandwidth and attenuators, calibrated antennas, etc...which are far beyond the scope of this book. But there are a few simple recommendations about EMI noise which should be considered when making bench measurements.

9.1 Ground Plane (Fig. 9.1)

First, if the objective is to make EMI free measurements, or to make ripple-noise evaluations without misleading results, due to other ambient noise, use a good ground plane on the surface of the work bench. This can be done by using aluminum metal foil (*Reynolds wrap* for instance) or a plain sheet of aluminum or copper. This will replace the undefined or noisy earth reference of the ac wall outlet. A paper overlay can be placed over the ground plane if there is a possibility of shorts with your test jig. This plane *will be the common ground reference for all of the test gear* and 0_{volts} will be tied to it by short braids or copper straps for:

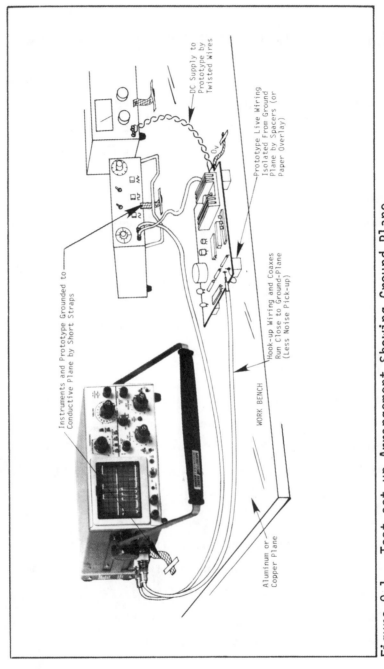

DC Supply to Prototype by Twisted Wires

Prototype Live Wiring Isolated From Ground Plane by Spacers (or Paper Overlay)

Instruments and Prototype Grounded to Conductive Plane by Short Straps

Hook-up Wiring and Coaxes Run Close to Ground-Plane (Less Noise Pick-up)

WORK BENCH

Aluminum or Copper Plane

Figure 9.1 – Test set-up Arrangement Showing Ground Plane.

- the chassis ground of the oscilloscope,
- the chassis grounds of all instruments
  (function generator, DVM, etc.),
- the ground of the device under test to
  avoid all possible ground loops.

The chassis of this equipment may be floated by
using a *cheater* on the power cords, but in this
case, the ground plane should be connected to the
earth ground for safety considerations.

9.2 Measuring Equipment

If an EMI receiver or spectrum analyzer is
not available, at least it is possible to make
some broad EMI diagnostics using an oscilloscope
if:

- it has a sufficient bandwidth: 20 MHz
  is an obsolute minimum and 100 MHz
  a good value for solving the most
  usual cases,
- the scope is entered via a 50Ω input
  impedance, if the signal in question
  can tolerate this loading. This les-
  sens the sensitivity but stabilizes
  the measurement in a constant impedance (remember a scope has its 1MΩ
  input shunted by $\simeq$ 20 pF of capacitance). See Fig. 9.2.

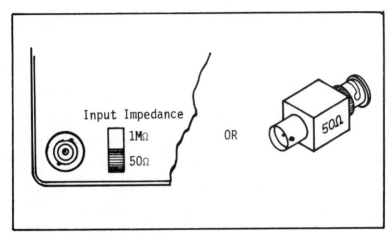

Figure 9.2 - Preferred Scope Input.

- If the signal or noise is greater
  than a few MHz, you must not use
  random wires, banana extensions
  and alligator clips and use only
  50Ω coaxes (Fig. 9.3)

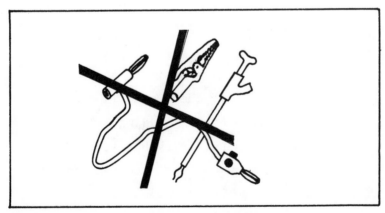

Figure 9.3 - Some Hook-up to Ban above few MHz.

The moderate sensitivity of a scope (usually 5 mV/div) can be enhanced by putting the 2 channels in series, in a dual trace oscilloscope. An auxilliary output of CHANNEL 1 is usually available, which, re-injected on CHANNEL 2, gives 1mV/div so that ≃ 200 μV can be detected (see Fig. 9.4).

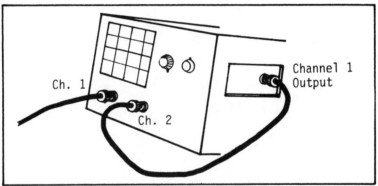

Figure 9.4 - Improving a 2-Channel Scope Sensitivity.

9.3 Voltage Probes

- Always use the shortest probe cable compatible with the set-up,
- Avoid lengthly ground clip leads (see Fig. 9.5),
- FET probes, or differential active probes are very efficient and will enhance the ability to make oscilloscope

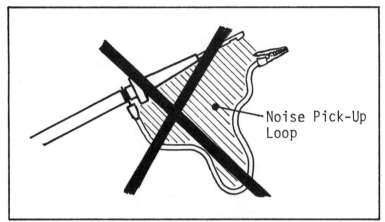

Figure 9.5 - Parasitic Noise Pick-Up by a too Long Ground Lead.

noise measurements. When a high impedance/low capacitance probe is used like the FET, beware of your body capacitance which can shunt the few picofarads of the probe by your *own* 100 to 200 picofarads, when touching the circuit under test.

9.4 Current Probes

Current probes are the most useful and versatile accessory in noise tracking. Regardless of the noise coupling mechanism (radiated or conducted, common or differential mode) *it will finally show up somewhere as a HF CURRENT* (see Fig. 9.6).

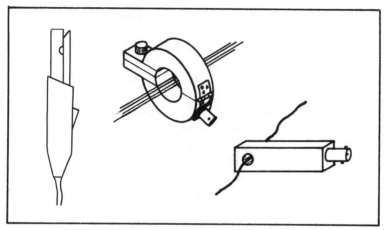

Figure 9.6 - Typical Current Probes.

When a probe is clamped on a whole cable harness (embracing incoming and returning signals), it will read only the CM current (the normal DM currents cancel the flux of each other). To use the probe, only the probe ratio must be known (called *transfer impedance*) in Volts/Ampere, i.e., how many millivolts the probe output will deliver for one milliampere flowing thru it. If, in an emergency mode, a current probe is not available, one can be improvised by using a few turns of wire around a magnetic or ferrite core. Of course, a rapid rough calibration must be made by placing a known current flowing in a passing-through wire at a few frequencies and measuring V_{out}, so that the probe ratio can be determined. It usually flattens and becomes constant above a few tens of kilohertz (kHz). See Fig. 9.7 below.

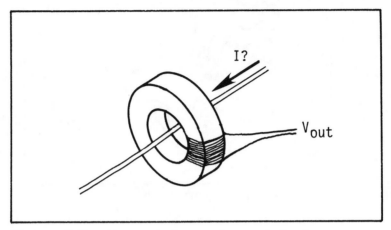

Figure 9.7 - Improvising a Current Probe.

9.5 Radiated Field Probe

Without a calibrated antenna and a tunable
sensitive receiver, real field strength measurement cannot be made, but two simple field sensors
can, at least, provide a qualitative *feel* and
give a rough numerical value of a radiated phenomena. These sensors are:

> ● *The Magnetic Loop Probe* - By winding
> a coil around an isolated mandrel,
> you can measure magnetic fields (see
> Fig. 9.8). To keep it from acting
> as an electric antenna, a Faraday
> shield is built in the form of a non-
> closed tube of copper or Al. The

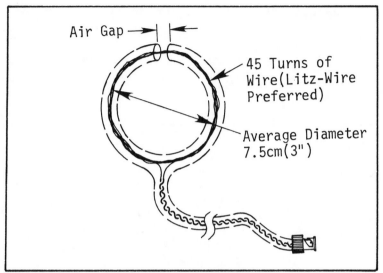

Figure 9.8 - Magnetic Loop Probe.

output of the probe is carried by a short shielded twisted pair, one lead terminating on the center pin of a BNC plug, the other lead attached to the outer shell (ground) along with the probe shield and the pair shield. For dimensions given, such a probe will deliver:

- 10mV per Gauss of field at 100 Hz,
- 100mV per Gauss of field at 1 kHz,
- 1 Volt per Gauss of field at 10 kHz.

This assumes that it is oriented to capture the maximum field (this is a very directional device).

- *The one meter (3 ft) rod antenna* (see Fig. 9.9).

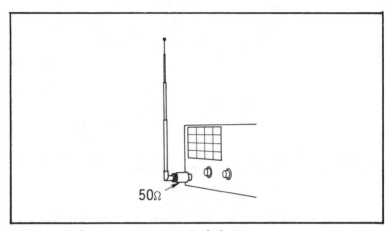

Figure 9.9 - One Meter Rod Antenna.

A one meter rod antenna, equipped with a BNC con-nector (the antenna terminates in the center pin) and followed by a 50Ω pad, can at least provide an idea of the ambient field, if it reaches the ordinary values for EMI concern (see par. 2.1). Such an antenna will deliver approximately:

- 3mV per Volt/meter of field around 1 MHz,

- 30mV per Volt/meter of field around 10 MHz,

● .3 Volts per volt/meter of field
around 75 MHz.

The ratio of $\dfrac{\text{E field in volts/m}}{V_{out} \text{ in volts}}$ is called the *antenna factor*. For instance, in Fig. 9.10, a 20 mV p-to-p out of an antenna at a frequency equal to $\dfrac{1}{2 \text{ x } .5\mu s}$ = 1 MHz is measured. The corresponding field is $1/2 \cdot \dfrac{20\text{mV p-to-p}}{3\text{mV/volt/meter}}$ = 6.6 volts/meter peak-to-peak, or 3.3 volt/m peak (2.3 volt/m RMS) which could possibly come from a powerful AM station located about 1 mile away.

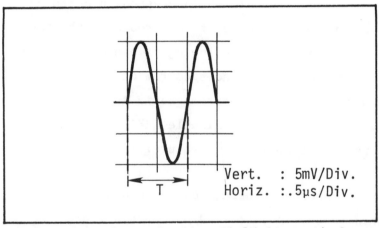

Figure 9.10 - Approximating Field Strength from a close AM Transmitter (≈ 1 MHz).

SUMMARY

While this book will not make a person a full fledged EMI Specialist, it will help one to avoid common pitfalls and misconceptions which can make a good design a total failure in the real and noisy environment of today.

If anyone should be interested in pursuing a better understanding of EMI problems, the handbooks listed on the following page are recommended. They cover a broad spectrum of information on how to control electromagnetic interference and would serve as a valuable addition to any library of technical literature.

As another option, contact Don White Consultants, Inc. to find out when a short course on interference control may be offered in your area.